BUGS! BUGS! BUGS!

FIRST EDITION
Project Editor Shaila Awan; **Art Editor** Susan Calver; **US Editor** Regina Kahney;
Pre-Production Producer Nadine King; **Producer** Sara Hu; **Picture Researcher** Martin Redfern;
Jacket Designer Natalie Godwin; **Publishing Manager** Bridget Giles; **Art Director** Martin Wilson;
Natural History Consultant Theresa Greenaway; **Reading Consultant** Linda Gambrell, PhD

THIS EDITION
Editorial Management by Oriel Square
Produced for DK by WonderLab Group LLC
Jennifer Emmett, Erica Green, Kate Hale, *Founders*

Editors Grace Hill Smith, Libby Romero, Michaela Weglinski;
Photography Editors Kelley Miller, Annette Kiesow, Nicole di Mella; **Managing Editor** Rachel Houghton;
Designers Project Design Company; **Researcher** Michelle Harris; **Copy Editor** Lori Merritt;
Indexer Connie Binder; **Proofreader** Larry Shea; **Reading Specialist** Dr. Jennifer Albro;
Curriculum Specialist Elaine Larson

Published in the United States by DK Publishing
1745 Broadway, 20th Floor, New York, NY 10019

Copyright © 2023 Dorling Kindersley Limited
DK, a Division of Penguin Random House LLC
23 24 25 26 10 9 8 7 6 5 4 3 2 1
001–333904–June/2023

A catalog record for this book
is available from the Library of Congress.
HC ISBN: 978-0-7440-7201-3
PB ISBN: 978-0-7440-7202-0

DK books are available at special discounts when purchased in bulk for sales promotions, premiums,
fundraising, or educational use. For details, contact: DK Publishing Special Markets,
1745 Broadway, 20th Floor, New York, NY 10019
SpecialSales@dk.com

Printed and bound in China

The publisher would like to thank the following for their kind permission to reproduce their images:
a=above; c=center; b=below; l=left; r=right; t=top; b/g=background

Dreamstime.com: Péter Gudella 19br, Martin Pelanek 11br, Salparadis 26, Sandra Standbridge 16crb;
Shutterstock.com: Sandra Standbridge 25tr

Cover images: *Front:* **Shutterstock.com:** Kenz_Hanson c, b/g, Lulus Budi Santoso br; *Spine:* **Shutterstock.com:** Lulus Budi Santoso

All other images © Dorling Kindersley
For more information see: www.dkimages.com

For the curious
www.dk.com

BUGS! BUGS! BUGS!

Jennifer Dussling

Contents

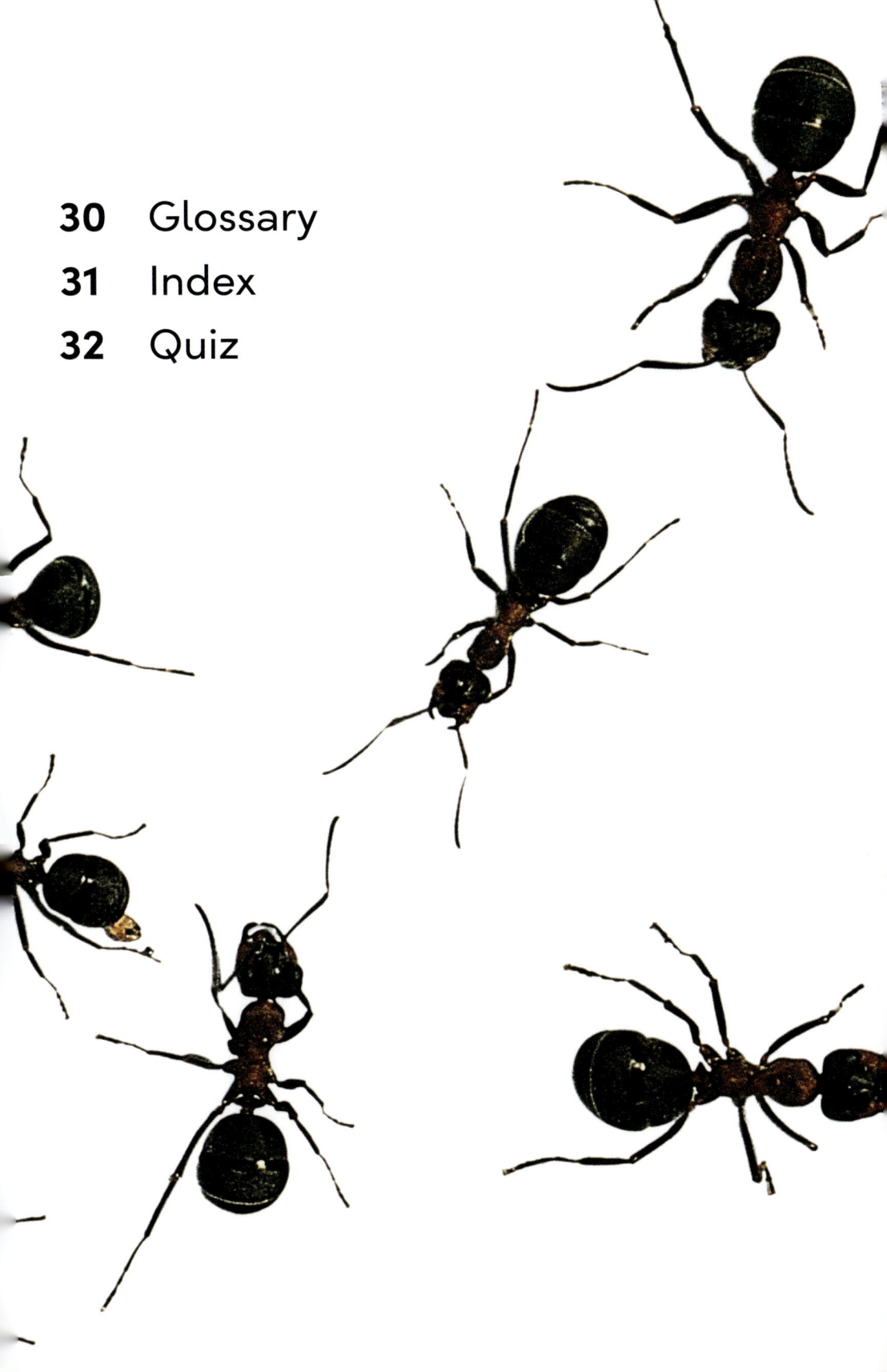

Bugs That Hunt

Yikes!
Bugs look scary close-up.
But you don't need to worry.
Most bugs are a danger only to other
insects. They are the bugs that really
bug other bugs.

Dragonfly

Stag beetle

Praying
mantis

Hunting
wasp

Praying Mantises

This praying mantis sits perfectly still.
But if you are a bug, watch out!

A fly lands on a branch near a praying
mantis. The mantis fixes its big eyes on
the fly. In a second, the mantis lashes out.
Its front legs trap the fly.

They pull it to the mantis's mouth.
Munch, crunch—soon the fly is gone!

Munch

Crunch

Hunting Wasps

Some bugs hunt other bugs, not for themselves, but to feed to their babies. This hunting wasp has just stung a beetle.

It will drag the beetle to its nest and lay eggs on the beetle. When the eggs hatch, the young wasps, called grubs, will eat the beetle up.

Hairy Food
One kind of wasp catches huge spiders for its grubs. It sometimes takes over the spider's home, too!

Dragonflies

It is a quiet day by a pond. One second, a mosquito is buzzing along. The next second, a dragonfly swoops down and snaps the mosquito right out of the air!

Ancient Insects
Dragonflies were around long before the dinosaurs! This dragonfly rotted away millions of years ago. It left its print in a rock.

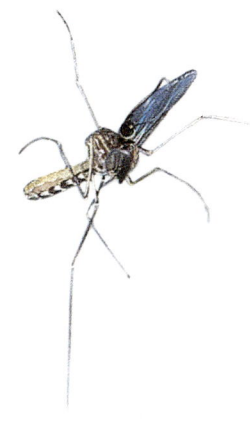

Dragonflies are flying killers that eat and eat and eat. In a day, they can eat their own body weight. That's like you eating 250 hot dogs!

Assassin Bugs

An assassin is a person who kills another person on purpose. The assassin bug is a bug that really lives up to its name.

Kissing Bugs
Some assassin bugs are called kissing bugs. That's because they often bite people on the face.

When it catches another insect, it
injects the insect with poison.
The poison turns the bug's insides to soup.
Then the assassin bug sucks up the soup!

Stag Beetles

Only one bug has to watch out for a male stag beetle—another male stag beetle! What do they fight about?
Usually a female stag beetle!

   The fighting beetles poke at each other, then lock jaws. One beetle grabs the other beetle and throws him. The loser scurries away.

Short, Sharp Jaws
A female stag beetle has smaller jaws than a male. But she can give a much sharper bite.

Bugs That Play Tricks

With so many killer bugs and other hungry animals, how do any insects survive? Some bugs have special ways to trick their enemies. Turn the page and read all about these bugs and more!

Postman butterfly caterpillar

Click beetle

Monarch butterfly
caterpillar

Stinkbug

Stinkbugs

Stinkbugs have glands that make smells. Some stinkbugs ooze a nasty-smelling liquid when they are in danger.

Stinkbugs are also known as shield bugs. Some use their flat bodies to shield their young from hungry insects and birds.

Monarch Butterflies

The monarch butterfly looks easy to capture and eat. But hungry bugs and birds leave it alone. Why?

In the insect world, bright colors are a warning. Bright orange signals that this butterfly tastes bad. Even the monarch caterpillars taste awful.

Changing Faces
When a caterpillar is fully grown, it changes into a butterfly inside a hard case like this, called a chrysalis (KRISS-uh-liss).

Caterpillars

Tropical lappet
moth caterpillar

This caterpillar's long hairs break easily.
When enemies try to catch it, they get a
mouthful of hair instead!

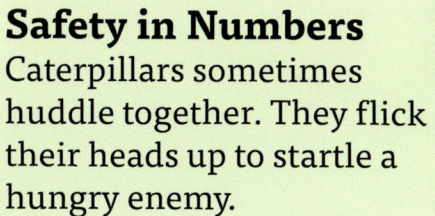

Safety in Numbers
Caterpillars sometimes huddle together. They flick their heads up to startle a hungry enemy.

Postman butterfly caterpillar

And this spiky caterpillar can be deadly. The leaves that it eats make its body poisonous. It is not harmed by the poison, but its enemies are!

Thorn Bugs

A thorn bug is good at hiding. It looks like a thorn on a twig. A bird looking for a meal might not see it.

Thorn bugs are smart,
too. They sometimes all
face the same way and
stay very still!

Jump!

Jump!

Click Beetles

To avoid being eaten, this click beetle has a clever way of escaping. It arches its back and then jumps into the air.

If the beetle lands upside down, it throws itself into the air again—this time hoping to land safely on its feet!

Flashing Lights
Some click beetles send out light signals. These flashing lights help the beetles to find a mate.

Glossary

Assassin bug
A bug with a narrow neck and long mouthparts, which it uses to stab its prey

Click beetle
A long, narrow insect that can "snap" its body with enough force to flip itself upright

Dragonfly
A flying insect with a long body and four horizontal wings

Hunting wasp
An insect that hunts and stings other insects and spiders

Praying mantis
A green or brown insect with a head shaped like a triangle and long front legs

Stag beetle
A brown and black insect with large, strong jaws that look like deer antlers

Stinkbug
A green or brown insect with a shield-shaped body that can release a bad-smelling liquid

Index

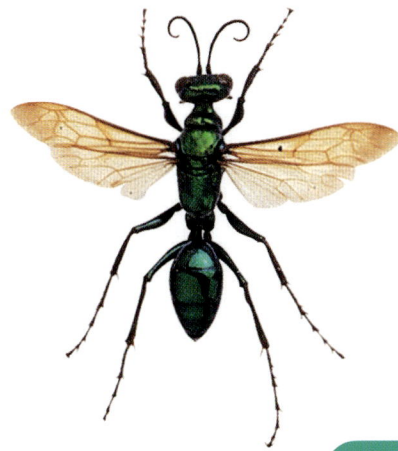

Quiz

Answer the questions to see what you have learned. Check your answers in the key below.

1. How does a praying mantis catch a fly?

2. Why does the hunting wasp sting a beetle?

3. How much can a dragonfly eat in a day?

4. Why are some assassin bugs called kissing bugs?

5. Which stag beetle has a sharper bite—a male or a female?

6. Why do some butterflies and caterpillars have bright colors?

7. How does a thorn bug hide from animals that want to eat it?

8. Why do some click beetles send out flashing light signals?

1. The praying mantis uses its front legs to trap it 2. So it can use the beetle to feed its babies 3. Its own body weight 4. Because they often bite people in the face 5. A female stag beetle 6. To signal to other bugs and birds that they taste bad 7. It stays still and looks like a thorn on a twig 8. To help them find a mate